THE **EXISTENCE** OF

GOD

STEVE KERN

For more information contact, Kern Enterprises,
2713 N. Sterling Ave., Oklahoma City. Ok. 73127
(405) 942-3504

All scripture references are from the New American
Standard Bible, translated by the Lockman
Foundation.

ISBN: 978-0-9798667-4-6

CONTENTS

FOREWORD

∞

There is nothing so critical to the human mind and spirit than the daily knowledge that God exists, that His word is true, and that He provides for my needs and my destiny. Here, we live in a space/time cocoon that prevents us from effectively seeing our past or our future, and our life experiences only leave a marginal residue that we call memory. It is in view of this that Solomon tells us to deliberately and continually rehearse, from our youth, facts that enforce and sustain the reality of God's creatorship, which he concludes, is the only way we can apprehend the purpose of our existence—to fear God and keep His commandments!

We are all steadily being exposed to alien data that is intended to desensitize us to the ever-present existence of a loving God. This clattering interference is ever attempting to dissuade us from His attending nearness. In this book, *The Existence of God*, Dr. Steve Kern is reminding us, through precept and example, of the many salient reasons for God's existence, His promises and His attention to my deepest requirements—now and in eternity.

Written is a comfortable manner, Dr. Kern presents arguments from prophecy, science, logic and the Bible that do not just suggest God's existence, but demand His existence! John the Apostle repeatedly reminded us that Jesus often said that He told us things that would come to pass, so that when they did come to pass, we would remember that He said them. In fact, John 13:19 tells us, *"Now I tell you before it comes, that when it is come to pass, you may believe that I am He."* (John 14: 29, 16:4) The reading of this brief treatise makes this conclusion overwhelming!

It is compelling that Apostle Peter wrote in 2 Peter 1:3 that we have been given all things pertaining to life and godliness through the *"knowledge"* of Him that called us. The words of the Holy Scripture are not randomly chosen; they are deliberate, and their meaning is also deliberate. This little word "knowledge" is introduced by a prefix that means exact or specific—a concept often used in medicine. Peter is not simply referring to a general knowledge that God has about us, but to a specific understanding about each of us. Salvation is not delivered in a "one size fits all" package. Salvation, while empowered by the shed blood of Jesus Christ, is distributed to every man and woman in a unique manner that is specially designed to suit the particular needs, make-up and personality that human being! Dr. Kern makes this fact explicitly clear.

Moreover, Kern makes it plain that something cannot come from nothing, that the universe is so complex and thoroughly designed until it does more than suggest the existence of God—it demands His existence! This is a must read for all hungry believers seeking more knowledge of the great Creator God.

- G. Thomas Sharp
Founder and President of Creation Truth Foundation

INTRODUCTION

∞

In 2008 a movie documentary titled **Expelled** featuring Ben Stein exposed how the concept of design was being stonewalled in higher education by removing educators who gave any hint that they might suggest that design arguments might have valid scientific support. The movie was very popular and opened the door to discussion about the weakening of evolution theory and a growing support for considering a higher intelligence as being the explanation for the existence of the universe. Since then more recent movies such as **God's Not Dead** and **God's Not Dead 2** have been released with large crowds in attendance. These and several other movies about faith in God show a growing interest in the pursuit of the existence of God.

Due to the influences of modernism's relativism of truth and postmodernism's denial of the existence of truth or at least our ability to know truth, more and more people are questioning the existence of God. Different expressions of atheism have become a fad in scholastic circles. The results has become a secularization of a large portion of western civilization that is now referred to as post Christian. Evolution has replaced creation as the ruling foundation of interpreting the existence of the universe in general and humanity specifically. During my time, since 1946, the existence of God has become considered by many as an old worn out relic of past cultural superstition.

Despite the many attempts to declare God dead and irrelevant, history is beginning to show once again that the eternal, self existent, Creator of the universe continues to refuse to die. Why is this so? There are several reasons. For one, each of us were created by God in such a way that causes us human beings to desire within ourselves something or someone beyond ourselves who we are an extension of. We call it a search for meaning. Furthermore, as we live on this earth and see its place in the universe, our logic and reason naturally speculates within our being that causes us to suspect that all that we see and experience could not just have happened by some random, chaotic, mistake. Consider also that there is a book made up of a bunch of books by different writers over a period of thousands of years that claims there is a God and gives a lot of physical evidence that what it claims is true. Interestingly, many who read this book develop a sense of knowing that this book is true. Many have tried to disprove its validity only to finally come to the conclusion that on the basis of their investigation the book has to be true. Of course the book I am referring to is the Bible. In it we have God revealing to us Himself and His purpose for creation.

Gene Cook Jr. gives an explanation of God's means of revelation that I have implemented in writing this book: "There are two different kinds of revelation, natural revelation, which we find in the world, and special revelation, which we find in the word of God. A basic rule of Biblical interpretation is that we should always interpret natural revelation through the grid of special revelation, that is, the word of God. We don't look at the world and then say this is the truth, therefore we must conform the Scriptures to what we see. This view is a deadly mistake and would lead to atheism.

I interpret science through the lens of the word of God, which presupposes the truth of Genesis 1–3. There is no evidence of higher value or authority than the word of God." In other words, Scripture is the truth. Logic, reason, science, observations of order are all subject to scripture authority without contradiction. These all only serve to verify what God has already revealed to be true.

I have written this book to help my readers find what I have found, that I have ample reason to believe that God does exist. I also want my readers to know that as I have chosen to believe in the existence of God, my choice continues to be reaffirmed on an ongoing basis. Jesus Christ gave a promise, *"Seek and you will find."* That has been my experience in seeking after the existence of God. As I have sought after God, I have learned that He was already seeking after me. Wow, God is good!

SCRIPTURES REVEAL
THE EXISTENCE OF GOD

In the beginning God created the heavens and the earth.
(Genesis 1: 1, NASB)

∞

The above quote is the first words recorded in the word of God we often refer to as the scriptures. These are very likely the first words ever written. Henry Morris in his book **The Genesis Record** believes them to be written by God as the introduction to His account of the creation week that He gave to Adam after Adam and Eve were created. They were then passed down from generation to generation until Moses used them as the beginning of his compilation of the book of Genesis.

This is the first statement of the **existence of God** we have in all of antiquity. Notice, this is a declaration of the existence of God, not an explanation. It is pure revelation. What I mean by revelation is that the verse gives us information about God that we would not know unless He Himself has told us. We know God exists because He has told us in Genesis 1: 1 that He exists.

From its very beginning the Bible does not seek to prove the existence of God, it accepts the existence of God as a fact. The proof of God's existence is understood to be found in what He has created. For example Psalm 19: 1 says, *"The heavens are telling of the glory of God; and their expanse is declaring the work of His hands."* (NASB) Paul wrote to the Romans, *"For since the creation of the world His invisible attributes, His eternal power and divine nature, have been clearly seen, being understood through what has been made, so that they are without excuse."* (Romans 1: 20, NASB) The creation is God's proof of His existence.

Just as the creation gives us proof about God's existence, the scriptures, in harmony with the creation, give us a progressive revelation of who God is and the kind of God He is. For example we learn of God's justice in Genesis 3 when He deals with the serpent, Eve, and Adam after they ate of the forbidden fruit. The serpent was cursed because it was used by Satan to deceive Eve. Eve was cursed because she listened to the serpent rather than obey God's command. Adam was cursed because he listened to his wife in disobedience to God.

Not only do we learn of God's justice in Genesis 3, we also learn of His grace when He gives His promise to break the curses of the woman and man through the seed of the woman. He also sacrificed animals to cover Adam and Eve's nakedness before they were expelled from the garden to live among thorns and thistles. God gave them leather for protection.

Through God's dealing with Abraham, Isaac, Jacob, and the nation of Israel, then Judah, scripture reveals God's love, His patience; that He is a person with a plan and purpose not only as the Creator but as a Father. Jeremiah 29:11, tells us, *"For I know the plans that I have for you, declares the Lord, plans for welfare and not for calamity to give you a future and a hope."* (NASB) We learn from scripture that God is a God of balance, of justice, and mercy. He punishes sin but gives mercy to those who humble themselves in repentance while asking for forgiveness.

Prophecy

One of the great revelations about God the Bible reveals to us is God's timelessness. Genesis 1:1 begins with *"In the beginning God..."* This statement tells us that God is preexistent to time and so is not limited by time but instead is the Creator of time. The Bible actually gives us proof of this fact through its use of prophecy. Isaiah 46:9-10 makes this point, *"Remember the former things long past, for I am God, and there is no other; declaring the end from the beginning and from ancient times things which have not been done, saying, 'My purpose will be established, and I will accomplish all My good pleasure...'"* (NASB) in order for God to know what will happen at the end of time, He cannot be limited by time. He must exist outside of time.

The Bible is full of examples of God's predictions of future events. In Genesis God warns Noah to prepare for the coming worldwide flood. He tells Abraham that his descendents will spend 400 years in Egypt and then be delivered from slavery. In the prophets, God warns Israel and Judah of their coming expulsion from their land and going into captivity. Daniel 9 gives a perfect description of the time line from Daniel to the coming of the Messiah. There are over 300 prophecies in the

Old Testament that speak specifically to the life and ministry of the Messiah fulfilled in the life and death of Jesus Christ the Messiah.

In order for the scriptures to exist as they do expressing the ability of the God of the scriptures to predict accurately future events over and over again, the God of the scriptures has to exist. Fulfilled prophecy requires the existence of God unlimited by time.

Miracles

The miracles in scripture reveal the existence of God. One calls for the other. It is as C. S. Lewis once said, "If we admit God, must we admit miracles? Indeed, indeed, you have no security against it. That is the bargain."

What is a miracle? A miracle is a witnessed phenomenon that cannot be explained by natural, physical laws. They are events that require a being to exist who has the power to override natural, physical laws. When you read Genesis 1, the God revealed to us there is the source of all natural law and thus is sovereign over those laws. This means He has the authority and power to override those laws according to His will and purpose.

Whether you read about the miracles in the Old Testament or the New Testament, they are all described in the context of eye witness events. Do a study of the evidence for the Genesis flood or the Exodus and you will discover that there is ample archeological evidence that support their authenticity. Jesus as the Son of God speaks of the flood, the Exodus, and even Jonah being swallowed by a large fish. If Jesus is God, as He claimed to be, by saying things like *"I and the Father are one,"* or *"Before Abraham was, I am,"* He must be the ultimate witness to these and many other miraculous accounts given in scripture.

In the New Testament, we have eye witness testimonies in the gospels, the book of Acts, and in various epistles. All of these writings have been proven to be reliable historical records by outside sources of every kind. Several scholars over the centuries from the time of Christ to now have thoroughly investigated the New Testament books and have come to the conclusion they can be trusted to be authentic. One such author among many is Josh McDowell and his books, Evidence that Demands a Verdict, and More Than a Carpenter.

Miracles reveal the existence of God in scripture. Miracles in scripture continue to be confirmed by the fact that miracles continue to be confirmed today. Testimonies abound where God has intervened in history to change the course of events in behalf of people and nations. There are testimonies of healings, and even people being revived from the dead. I am not going to go into specific examples. I just want it to suffice to say miracles still happen and there are many sources you can look up to read about them for yourself. But know this, miracles continue to reveal the existence of God.

Jesus Christ

The greatest revelation of the existence of God found in scripture is the person of Jesus Christ. If Jesus Christ is who He claimed to be then He solidifies the fact that God does exist. There is no doubt that Jesus was a historical person. Given the testimonies of those who knew Him best, they record that He verbally claimed to be God, and they personally testify that they were convinced that He is God in a physical, human body. Each of them were committed to Jesus' deity to the point that most were martyred for their faith.

They claimed that an angel announced that He was *"God with us."* (Matthew 1: 23, NASB) They recorded His own words such as, *"I and the Father are one."* (John 10: 30, NASB) *"Before Abraham was, I am"* (John 8: 58, NASB) *"He who has seen Me has seen the Father."* (John 14: 9, NASB)

The apostles personally referred to Him as God. John wrote, *"In the beginning was the Word, and the word was with God, and the Word was God."* (John 1: 1, NASB) Peter wrote, *"...by the righteousness of our God and Savior, Jesus Christ."* (II Peter 1:1) Paul wrote, *"And He is before all things, and in Him all things hold together."* (Colossians 1: 17, NASB)

Once you become convinced that the New Testament can be trusted to be a compilation of historical eyewitness accounts of the life and ministry of the man Jesus Christ being God come in the flesh, you will never question the existence of God. Actually, the whole Bible is given by God as a historical record to reveal His existence. Jesus Christ is the central figure from beginning to end. That is why He refers to Himself in Revelation 22: 13, *"I am the Alpha and the Omega, the first and the last, the beginning and the end."* (NASB) Jesus talking to the religious leaders of His day also said, *"You search*

the scriptures, because you think that in them you have eternal life; and it is these that bear witness of Me..."
(John 5: 39, NASB)

The scriptures reveal the existence of God. Because Jesus Christ proved Himself to be God in the flesh through His miracles and resurrection, and the fact that He is a proven, recorded historical person; He is the scriptures greatest revelation of the existence of God.

LOGIC & REASON DEMAND THE EXISTENCE OF GOD

"The fool has said in his heart, 'There is no God'"
(Psalm 53: 1, NASB)

∞

Just as the scriptures reveal the existence of God, our experience of living in His creation as His creation forces us to conclude that God has to exist. A personal definition I have of a fool is someone who is looking at the facts before his very eyes and denies that they are there. The popular quote, "Don't confuse me with the facts" fits the fool perfectly. It is amazing to me that so many of those who see themselves as the true purveyors of logic and reason are often those who fit my definition of a fool best.

A quote from Winston Churchill fits well here: "Men stumble over the truth from time to time, but most pick themselves up and hurry off as if nothing happened."

Logic and Reason

What is logic and reason? Where do they come from? I will answer the first question first. R. C. Sproul in his book, **Defending Your Faith**, speaking of Aristotle's description of logic writes, "But we must keep in mind that Aristotle did not invent logic; rather he defined it. He argued that logic is a necessary tool for human thinking and communication, as well as a means for us to comprehend the rational structure of the universe." (p. 36) Logic then is a set of laws that are not developed by the thoughts of men; instead they are discovered as being innate to the makeup of the creation. These laws of logic reveal the attribute of the Creator as being logical and the Absolute who established reason. The God of the Bible is a God of reason who established the laws of logic.

Reason is "The ability to think, form judgments, draw conclusions, sound thought or judgment, good sense; to think coherently and logically; draw inferences or conclusions from facts known or assumed." (Webster's

New World Dictionary) In other words, reason is the mind thinking while being led by the laws of logic that leads to right understanding. Not all thinking is reasonable. Thinking is only reasonable when it is logical.

Now that we know what logic and reason are, I want to answer the second question, "Where do logic and reason come from?" The Bible tells us in Genesis 1: 26 that God made us in His own image. The fact that we can think logically based on reason makes us conclude that God Himself is the absolute cause of the laws of logic that allows us to think reasonably just as He does. This being true it stands to reason that God gave us logic and reason as a means of knowing Him and having the ability to discover how His creation works.

Galileo based his whole life of scientific investigation on logic and reason. He is quoted as once saying, "I do not feel obligated to believe that the same God who endowed us with sense (logic), reason, and intellect had intended for us to forgo their use." Galileo understood that logic and reason could only come from a God who has the power to endow His creation, made in His image, with these same qualities He expresses infinitely.

The Necessity

Now the title to this chapter is **Logic and Reason Demand the Existence of God**. What do I mean by that statement? The laws of logic joined together with reason lead to one conclusion. God has to exist. Actually, in the study of Theology, Thomas Aquinas developed five reasons based on logic for the existence of God. One of those reasons is the **necessity** of God. Through arguments based on laws of logic, Aquinas reasoned that according to the reality of the being or existence of the universe an absolute Being has to exist. Saint Anselm in the eleventh century was the first to articulate this argument that Aquinas later in the sixteenth century labeled the **Ontological argument**. The Greek word *onto* means "being" and thus the study of the being of God.

This argument reasons that if there was ever nothing there would always be nothing because nothing has **no** ability to produce something. On the other hand if there is something then there has always had to be something in existence knowing that nothing could never produce something. Because there is the something we call **being**, it is logical to conclude that there must be an ultimate **Being** out of which all other things that **be** must come. I can say it more succinctly by saying **because we are God is** or **because God is we are**. There can be no other conclusion.

That brings us to the law of logic referred to as **cause and effect**. Frank Turek in his book **Stealing From God** states how important this law is, "To doubt the law of causality is to doubt virtually everything we know about reality, including our ability to reason and do science. All arguments, all thinking, all science, and all aspects of life depend on the law of causality."

What does this law say? Every effect requires a cause that is greater than the effect. To understand this statement you need to understand what is meant by an effect. An effect is anything that cannot exist by its own doing. It requires something outside of itself that has the ability to bring the effect into existence. That something is a **necessity** we call a **cause**. Look around you. What do you see? You see all kinds of effects that exist because something or someone caused them to exist. Your car, your cell phone, your house, an organized garden, they all exist because they have a cause with the ability to plan them, make their parts, and organize them into their final form.

The universe itself is an effect. It is not self existent. If it were it would not have had a beginning. If it had a beginning then there would have to have been a time when it did not exist, meaning it had to have been nothing before it began to exist. Logic has already determined that nothing will always remain as nothing. So because the universe is something that had a beginning we must conclude that it had to have been brought into existence by a pre-existent Cause greater than itself.

This cause and effect fact excludes atheistic evolution as an explanation for the existence of the universe. Why? Because it is a theory that tries to explain an effect without providing an acceptable cause. The most popular explanation of this theory suggests that the elements of the universe were compressed into the size of a small dot called a **singularity**. These elements had to be inorganic, lifeless matter. This singularity also had to always exist. At some point the singularity exploded and began to expand into what now has become the universe.

This theory creates a lot more questions than it answers. For one the question has to be asked if the universe is all that exists, then where did the singularity exist outside of itself before it exploded? Then you have to ask where the energy came from to cause the singularity to explode? Energy is not self existent. It has to have a source. The law of motion tells us stationary objects tend to stay stationary. That law begs the question, what moved within the singularity after its infinite existence to cause it to explode?

Then the questions come as we observe effects that the singularity provides no explanation for in our life experience. Questions like: Where did life come from? Where did intelligence come from? How did a chaotic explosion provide all the design and order we observe? Why is there morality? What causes emotions, etc.? The singularity provides no answers for these questions.

In the study of religions, the God of the Bible is the only one that establishes the existence of the Being necessary to exist in order to provide the reasonable Cause to answer all the above questions and many others. The God of the Bible is a living God. He is the eternal Being preexistent to time. He is a personal, moral, and emotional God. He is the God of the three big "O's," omnipotent (all powerful), omniscient (all knowing), and omnipresent (everywhere at the same time). He is the absolute truth who upholds all that exists.

Sir Isaac Newton was born in 1642 and died in 1727. He was known mainly for his discoveries in math and science and yet much of his studies in these disciplines were used by him to develop his theology. He was a believer in the God of the Bible and much of his faith was undergirded by what he learned from the creation in relationship to the revealed word of God. Newton wrote in his **Principia, 1687**, "The most beautiful system of the sun, planets, and comets, could only proceed from the counsel and dominion of an intelligent and powerful Being... All variety of created objects which represent order and life in the universe could happen only by the willful reasoning of its original Creator, whom I call 'Lord God' ... This Being governs all things, not as the soul of the world, but as Lord over all; and on account of His dominion He is wont to be called 'Lord God'... The supreme God exists necessarily, and by the same necessity He exists always and everywhere." Obviously Isaac Newton understood logic and reason that require the **Necessity**...

The arguments I have given you here are fundamental to logic and reason. There are many books that go into far greater detail. The point is that logic and reason do not just suggest the existence of God, they demand the existence of God. That is why He is referred to as **the**

Necessity. Without the God of the Bible there is no logical reason for anything to exist.

SCIENCE & OBSERVATION CONFIRM THE EXISTENCE OF GOD

The heavens are telling of the glory of God;
and their expanse is declaring the work of His hands.
(Psalm 19: 1, NASB)

∞

Science is all about observation. It also requires making logical conclusions about what we observe. The Bible tells us that observing God's creation is one way we can confirm the existence of God. Paul wrote to the Christians in Rome, *"For since the creation of the world His invisible attributes, His eternal power and divine nature, have been clearly seen, being understood through what has been made, so that they (ungodly people) are without excuse."* (Romans 1: 20, NASB)

The principles of scientific study were developed early on by men who were mainly Christians. Francis Bacon

(1561-1626), a member of the Church of England, is considered to be the father of the scientific method. The reason why they did scientific studies was in their words, "To think God's thoughts after Him." They believed that the more they could learn about God's creation the more they could learn about God. Things like who God is, and how He thinks, what the extent of His power is, what the physical laws are that He established to make the physical universe work, and many more. Scientist James Tour made the point, "Only a rookie who knows nothing about science would say science takes away from faith. If you really study science, it will bring you closer to God."

The early Christian scientists believed that they had a mandate from God to do their scientific studies. After God created Adam and Eve, He said to them, *"Be fruitful and multiply, and fill the earth, and subdue it…"* (Genesis 1: 28, NASB) The Hebrew word for **subdue** is **kabash** and means to conquer, subjugate, or bring into subjection. That is what science is all about; learning how the creation works and then bringing it under control for good use. So God gave us science so we could know Him through His creation that confirms His existence.

What are some of the findings in the study of science that have confirmed the existence of God? I will share three examples: the study of quantum mechanics; the necessary make up of the creation that makes science possible; and the age of the earth.

Quantum Mechanics

I don't claim to be an expert in the area of quantum mechanics, but there are some interesting observations that have surfaced in the study of particle physics that I can share. Unlike the predictability of the actions of objects in the observable universe reacting within the laws of physics, the particles of atoms are not so predictable. The observation of the unpredictable movements within the subatomic realm such as particles that are not observably subject to time suggest that these particles are moving through other dimensions beyond the physical dimensions of time, space, and matter. There are different theories that try to explain this phenomenon but the fact remains that subatomic particle studies suggest the existence of **other dimensions**.

How does that confirm the existence of God? Let's go back to Genesis 1: 1, "In the beginning God..."

That statement declares that God as the Creator of the physical dimensions of time, space, and matter was preexistent to those dimensions. That fact tells us that before the universe was created God existed and still exists in dimensions beyond the universe. The Bible refers to these dimensions as **the spiritual realm**. So then it is reasonable to assume that there are other dimensions beyond time, space, and matter. Actually what Genesis 1: 1 does is predict that science would eventually discover that there are other dimensions. Quantum mechanics supports that prediction.

It stands to reason that according to the law of cause and effect we have to conclude that energy requires a Cause. Colossians 1: 17 tells us that Jesus, as God in the flesh, is the source of that energy when it says, "And He is before all things, and in Him all things hold together." Einstein's formula, E=MC squared, tells us God is the source of energy. His being the light energy suggests there would be interaction with matter in the dimensions where God dwells. E =energy, M=matter or mass, C squared=the speed of light squared. In other words matter is dependent on light energy to hold its particles together. Particle physics has discovered subatomic particles must be interacting with dimensions from God's existence in which they derive and sustain their energy.

This is important to know because quantum mechanics blows the lid off of the naturalist scientist's assumption that the universe is all there is and nothing exists outside of time, space, and matter. The Bible makes it clear that there must be innumerable dimensions beyond this created, limited, universe providing an infinite abode for the existence of an infinite God not limited by His created time, space, and matter. Science has begun to make that clear as well.

Science Requires Absolutes

Peter Huff once stated, "The atheist can appeal to nothing absolute, nothing objectively true for all people, it is just mere opinion enforced by might. The Christian appeals to a standard outside himself/herself in which truth and qualitative values can be made sense of." This quote was given in reference specifically to moral values and yet the principle of a necessary absolute applies just the same to the laws of science.

Dr. Jason Lisle has a wonderful book titled, **The Ultimate Proof of Creation**, that exposes how naturalist scientists depend on absolutes to do their scientific investigations. They do so even though their atheistic worldview denies the necessary Source for the absolutes they depend on. Dr. Lisle states the summary of his book in this way, "The ultimate proof of creation is this: if biblical creation were not true, we could not know anything." (p. 40) His point is that to know anything to be true for sure there must be a reason to believe that what we observe can be trusted to always be true.

The existence of the God of the Bible provides the absolutes necessary to know there are some things that are always true. This would include all the natural laws science investigation depends on to come to solid knowable conclusions. A universe that began by a chaotic explosion that was formulated by unpredictable, random, events has no reason to justify that anything can be depended on to always be the same. The way things are now may not have been the way they were in the past nor may they be that way in the future. This is true because chaos can only produce chaos.

The principle of conformity (predictability) means that when you do an experiment following the exact same steps, and using all the exact same ingredients you will always get the exact same result. That being true then there has to be an absolute cause of conformity behind the whole process. The Bible gives us that absolute when it tells us that God (Jesus Christ) *"... is the same yesterday, and today, and forever."* (Hebrews 13: 8, NASB) Malachi 3: 6 states, *"For I, the Lord, do not change..."* (NASB) God assured Noah that things would always be the same in what He has established, *"While the earth remains, seedtime and harvest, and cold and heat, and summer and winter, and day and night shall not cease."* (Genesis 8: 22, NASB) God as the absolute, unchanging, Creator of the laws of nature provides the reason why we can trust the principle of conformity. Because God does not change, we can trust His natural laws will not change either.

The Age of the Earth

The big bang and evolution process used to describe how everything in the universe came into being requires multiple billions of years of development to make the process feasible. A literal interpretation of the creation week and the subsequent chronology of human history given in scripture only allows for 6,000 to 10,000 years. If it can be shown using science that the earth is only several thousand years old and not billions, a creator God would become an absolute necessity. There would not be nearly enough time for a Godless evolution to work. So, if the Bible is true then there should be observable evidence that suggests the earth is less than 10,000 years old.

There are hundreds of books that give documented evidence that the earth is very young. Three observations from science covered in some of these books are: magnetic field decay, C14 in coal, diamonds and dinosaurs, and helium in zircon crystals. I will take each one, one at a time.

Magnetic Field Decay

Some of the most documented and observable laws of nature are the laws of thermodynamics. The second law is the law of entropy or better known as the law of decay. Dr. Henry Morris describes this law in his book **Scientific Creationism** as, "...a universal change in nature which is downhill..." (p.38) This means that the different forms of energy are in a continual state of decay. Energy is not increasing, it is decreasing. One example would be the sun. The sun is not getting hotter, it is in the process of burning out like all the other stars.

Like the sun, this process of decay applies to the earth's magnetic field. Dr. Thomas Barnes, a onetime professor of Physics at the University of Texas at El Paso, did a study of the magnetic field using 135 years of records that recorded the strength of the magnetic field during that time. On the basis of analytical and statistical studies Dr. Barnes was able to conclude that the magnetic field has a decay rate with a half life of 1400 years. That means that if you go back in historical time every 1400 years the strength of the magnetic field would be double what it was 1400 years in the future.

If we do the calculations, we will discover that the magnetic field in just 10,000 years would be as strong as a magnetic star. Here is the problem. The earth does not have the thermonuclear processes to produce a magnetic field of that kind of strength. This fact must lead us to conclude that the magnetic field must have been operating at maximum strength just a few thousand years ago.

The literal biblical chronology only allows the earth age to be a little more than 6,000 years if that long, depending on which interpretation you follow. We know the curse of death took place in Genesis 3 around a hundred years after the creation. The creation was pronounced by God as being very good suggesting everything was working at a perfect optimum level. The curse of death not only applied to Adam and Eve, it also applied to all the rest of creation. This apparently was when the law of entropy or decay was established and everything in the creation, including the magnetic field, began to die.

This time frame fits very well with what we know about the decay of the magnetic field. About 6,000 years ago it was at full strength and since that time it has decayed to the level of strength that it is at now. This fact begs the question "What was the environment like back when the magnetic field was far stronger than it is now?" That is a topic for another book but you can get some ideas about the answer in my book, No Other Gods.

The point is that the scientific studies of the magnetic field do not allow the earth to be more than a few thousand years old. That being true, these findings do not allow nearly enough time for evolution to work. That being the case, only the God of the Bible introduced in Genesis 1 can be the plausible reason behind the existence of the universe.

The study of the earth's magnetic field gives strong confirmation from science that God exists and Genesis 1 is literally true. C. S Lewis once quipped,

"No philosophical theory which I have yet come across is a radical improvement on the words of Genesis, that 'In the beginning God made Heaven and Earth.'"

C14 in Coal, Diamonds, and Dinosaurs

The carbon dating method referred to as C14 is used to date how **long ago something that was alive died. Carbon 14 is an unstable isotope produced in the upper atmosphere of the earth as radiation from the sun enters the atmosphere and bombards Nitrogen-14 transforming it into radioactive Carbon 14. The Carbon 14 then begins to decay back to Nitrogen-14 through the beta-decay process. The Carbon 14, like the non-radioactive Carbon 12 in the atmosphere is absorbed into the physical bodies of living organisms. When that organism, whether plant or animal, dies it no longer absorbs the Carbon 14 or Carbon 12. The Carbon 14 in the dead organism continues to decay at a rate of a half life of 5370 years.**

What does all that have to do with confirming a young age of the earth you might ask? Well, it has been estimated that after a period of around 29,000 years, there should no longer be enough traces of C14 left in a fossil to give any possible reliable date as to when the organism died. It has also been determined that dates determined by the C14 method can only be considered reliable up to 5,000 years give or take a thousand. But my point is that C14 should be totally gone out of an organism's fossil remains after no more than 80,000 years, some say 100,000 years at the most.

Here is the problem for those who want the earth to be 4.5 billion years old. There have been extensive studies that have discovered C14 in objects that are supposed to be billions of years old. Some examples are coal, diamonds, and dinosaurs. The **Institute for Creation Research** (ICR) published findings from their RATE studies ending in 2005. Their studies involved several top scientists in their fields of study who followed extensive research protocol. Dr. Don DeYoung published a summarized report titled **Thousands not Billions** that gives concrete evidence of C14 still remaining in coal and diamonds. When adding the change in the amount of less C14 in the atmosphere of the pre-flood environment they were able to determine the dates of the coal and diamonds studied to be only 5,000 years ago which fits the biblical time frame very well.

Since that time, ICR has done further studies on C14. They have found C14 in dinosaur bones that are supposed to be 65 to 75 million years old. In 2015 spring addition, Creation Research Society Quarterly published an article on research done on several fossils including dinosaurs in different levels of prehistoric rock layers. The article states, "Five different commercial and academic laboratories detected carbon-14 in all samples, whether Cenozoic, Mesozoic, or Paleozoic source rocks." This suggests that these rock layers and the fossils in them were buried at the same time. They are better explained by the flood of Noah's day than billions of years of slow gradual layering. The scientific evidence keeps getting stronger and stronger that supports a young earth date. If the earth is young, that fact confirms the God of the Bible has to exist.

Helium in Zircon Crystals

Zircon crystals are not found in abundance. They are believed to be some of the oldest minerals on earth. They are very hard and not very porous. There is a certain amount of uranium decaying in the zircons that produces helium.

Helium atoms are very small and do not combine with other atoms to form larger molecules. As a gas, helium atoms are in constant movement. All this being true, even though zircons are very hard and tightly bound, helium still has the capacity to escape out of the crystals once they have been formed.

Here is the problem. Even though the zircons are tightly formed the helium in them can still escape out of them at a certain rate. That being the case, if the zircons are as old as accepted dating processes say they are, then there should be little or no helium left in them by now. That has been found to not be the case. As a matter of fact the helium content in the zircons is very high.

The ICR RATE studies on zircon crystals discovered, "...helium diffusion measurements show that such high concentrations of helium simply cannot be sustained for more than a few thousand years." (Thousand not Billions, p. 78) If the zircons are as old as old earth science says they are, then all the helium should be gone. And yet, helium remains in the zircons at a high rate. Here again science confirms a short age of the earth which requires the existence of the God of the Bible.

There are many other ways that science confirms the existence of God. One of the best is how scripture speaks of many truths about the creation that we are only just now beginning to discover. I have already used Genesis 1: 1 that establishes the existence of other dimensions where God exists beyond the creation. Psalm 104: 2 speaks of an expanding universe predicted by Einstein and proved by Hubble where it says that God stretches out the heavens like a curtain. Job 26: 10 declared that the earth was round long before Columbus set out to prove it in 1492. The verse says, "He has inscribed a circle on the surface of the waters..." (NASB)

The only honest reason anyone who has chosen to not believe in the existence of God is because that person chooses not to believe. The Christian faith is an objective faith. An object faith is based on being able to make observations that lead to reasonable conclusions. We can observe fulfilled prophecy in the Bible historically. We can observe the fact that Jesus Christ actually lived and hundreds of people witnessed His resurrection. We can observe the reality of effects like life, intelligence, design, and morals that require a Cause. We can observe that science requires absolutes in order to know discoveries are facts. Faith in Jesus Christ is not something we hope to be true, but something we can know to be true, both by physical and spiritual insight. It is a joining of the mind and heart directed by the Holy Spirit.

THE HUMAN SOUL LONGS FOR THE EXISTENCE OF GOD

I stretch out my hands to Thee; my soul longs for Thee in a parched land. (Psalm 143: 6, NASB)

∞

Woody Allen, the comedian and film director is quoted as having said, "The artist's job is not to succumb to despair but to find an antidote for the emptiness of existence."

C. S. Lewis in his testimony told of his struggle as an atheist with his own inner turmoil, "I was at this time of living, like so many Atheists or Anti-theists, in a whirl of contradictions. I maintained that God did not exist. I was also angry with God for not existing. I was equally angry with Him for creating the world."

Then there is the famous quote attributed to Blaise Pascal that is really an interpretation of the original quote, and yet is still powerfully true, "There is a God-shaped vacuum in the heart of every person, and it can never be filled by any created thing. It can only be filled by God, made known through Jesus Christ."

The Longing Soul

I believe this longing of the human soul for God is the most powerful argument for the existence of God there is. This longing is universal. It is the reason why man is such a religious being. There is this desperate search for meaning that every human looks for; and if this meaning is not found this empty search leads to despair. The Woody Allen quote above is a perfect example of this fact. Having meaning for our existence is the antidote for the emptiness he speaks of.

The book of Ecclesiastes is the testimony of King Solomon telling how he tried everything in life to discover the meaning of life. He starts by declaring, *"Vanity, of vanities! All is vanity."* (verse 1: 2, NASB) He then concludes, *"I have seen all the works which have been done under the sun, and behold, all is vanity and striving after wind."* (verse 1: 14, NASB) In chapter

46

9: 10 he suggests, *"Whatever your hand finds to do, verily, do it with all your might; for there is no activity or planning or wisdom in Sheol where you are going."* The Preacher, as Solomon refers to himself, goes on to say in 9:12, *" Moreover, man does not know his time: like a fish caught in a treacherous net, and birds trapped in a snare, so the sons of men are ensnared at an evil time when it suddenly falls on them."*

The important thing to understand about Solomon's cries of meaningless **vanity** is that he leaves God out of the equation until the final chapter of his testimony. He states in his final chapter 12: 1 and 13, *"Remember also your Creator in the days of your youth before the evil days come and the years draw near when you will say, 'I have no delight in them'... The conclusion, when all has been heard, is: fear God and keep His commandments, because this applies to every person."* Solomon concluded that life is vanity or meaningless when our Creator is left out of our lives.

Empty Stuff

Have you ever wondered why some people who seem to have it all when it comes to possessing the things of this world commit suicide? They have fame, wealth, power, lots of friends, and yet none of it brought a sense of fulfillment within themselves. The meaninglessness of their empty lives led to despair, and the despair led to ending it all. In doing so, they became fulfillments of what Jesus exposed as Satan's goal of his diabolical deceptions when He said, *"The thief comes to steal, and kill, and destroy..."*

Satan often uses **stuff** to make us think we can find meaning in life through the **stuff** we try to fill the void in our lives with. The **stuff** can be worldly goods, all kinds of pleasure, adventure, new kinds of highs, and yet when all is said and done we are still left with only our empty self alone. The longing of the soul only grows stronger.

There are so many tragic figures who seemed to have it all and yet took their own lives. Ernest Hemingway, Marilyn Monroe, Jim Morrison, Kurt Cobain, Robin Williams just to name a few. They are all casualties of the spiritual war we find ourselves in; victims of Satan's deceptions. I believe having an increase in having it all actually heightens an awareness of futility within a person. They are surrounded by everything that is supposed to satisfy and yet none of it does. Those who do not have it all can still hope having more will be their deliverance, but when those who do have it all come to a dead end thinking there is no place else to turn, that is when despair sets in.

What Satan works to do is keep a person from learning that there is some place else to turn. After Jesus exposed Satan's objective to destroy us, He went on to say, "...I came that they might have life, and might have it abundantly." (John 10: 10, NASB) Jesus here is declaring that He, as God, can make an empty life full with meaning and make it worth living. He can do that because in Him we find our Creator; and in Him we find a meaningful eternity ahead of us that will take us beyond the vanity of this fallen world. In Jesus Christ we find our reason to exist. Jesus came to satisfy our longing for God when He promised, "Come to Me, all who are weary and heavy-laden, and I will give you rest." (Matthew 11: 28, NASB)

The Source of a Longing Soul

We have to ask ourselves, "Where does this longing for the existence of God come from?" For the answer we need to go back to our discussion of cause and effect. The human soul longs for God because we were created by God in His image. We were meant to know Him and have fellowship with Him, living in His presence. God is the cause of our longing. Our longing soul is an effect caused by God when He removed His presence from us because of the sin of Adam and Eve that has passed on to us all. We can only find meaning in life when we learn that our purpose for existing is God and start to live in that purpose in a restored relationship with God. Those who do this experience Psalm 107: 9, *"For He satisfied the thirsty soul, and the hungry soul He has filled with what is good."* (NASB) God is good and satisfies the longing of the human soul as only He can with His presence living in us.

I said at the beginning of this chapter that I believe the longing of the human soul is the greatest evidence for the existence of God. Here is my major reason why I feel this way. The other three reasons I have given for the existence of God: God is revealed in scripture, logic and reason demand the existence of God, and Science and observation confirm the existence of God, each can be considered with the mind as objective realities that give us reasons to believe in God. But, the longing of each of our souls is something we experience in the depths of our being. As Psalm 42: 7 states, *"Deep calls to deep..."* (NASB) From the depths of our despair we cry out to God in faith and find He is deep enough in Himself to fill the depths of our souls. When we enter into a personal relationship with Him through Jesus Christ, that longing is replaced with meaning. The whole experience becomes a part of our being that is undeniable. The experience expands our faith from our minds into our being. Our satisfied soul becomes who we are in experience, and that transcends anything we can know with our minds.

An Honest Prayer

Several years ago I had a friend who was an agnostic. He had attended church as a young boy but became disillusioned with religion in college. When I meant him he was finishing a degree in accounting in order to take the CPA test in Texas. To get enough credits he had to gain credit hours in P.E. The college he was attending agreed to let him achieve those hours by playing tennis for so many hours during his last semester. So he asked me to play tennis with him so he could finish his degree. I had played a little tennis so I agreed.

After we were into our agreement for several weeks, knowing I was a pastor, he finally opened up to me and stated out of the clear blue, "I have a problem with this thing called faith. I just don't get it." So I asked him, "Steve, if God exists would you want to know him?" He paused for a moment and then he said, "Yes, I would want to know Him." So I said, "Ok, tell Him" With a puzzled look on his face he asked, "What do you mean?" "Just say it out loud, God if you exist, I want to know you." So as honestly as he knew how, as we stood there across from each other at the tennis net, he said out loud, "God if you exist, I want to know you."

We went on and finished our match and that was it. I didn't say another thing to him about his reaching out to God. It was about two weeks later that Steve gave me a call and wanted me to come to his house, and so I did. When I walked into his living room he said, "He did it." "Who did what," I asked? "God has taken my doubts away, and I want to give my life to Jesus Christ." He and his wife both prayed to receive Christ that night and became active servants for God for the rest of their lives.

I want you to notice that I didn't tell you how God answered Steve when he reached out to Him. I didn't do so because God reveals Himself in ways that works for different individuals in different ways. If you have a desire to know God, and are willing to call out to Him, God's answer for you will not be how He answered Steve. But, however He answers, He will answer. My point in sharing this experience is that God will answer the cry of an honest heart that truly desires to know Him. It was true for me. It was true for Steve. It will be true for you, if you honestly want to know God.

I can assure you that God does exist. I encourage you to believe in Him. When you do, you will come to know Him. The way faith works with God is believing leads to knowing. As you come to believe in Him you will come to know Him more and more. Every person's experience will be different but it will be very real to that person God reveals Himself to. Is that person you?

CONCLUSION

I am the Alpha and the Omega,
the first and the last, the beginning and the end.
(Jesus Christ, Revelation 22: 13, NASB)

∞

"I believe in Christianity as I believe that the Sun has risen, not only because I see it, but because by it I see everything else. " C. S. Lewis through this quote has described my own personal experience in developing my Christian biblical worldview. When I came to understand that the Old Testament prepares the way for the coming of the Messiah, and that the New Testament reveals the fulfillment of the Messianic Old Testament in the person of Jesus Christ, the reality of what my existence is all about became very clear. Coupled with scripture, logic, and science, special revelation (scripture) and natural revelation (creation), each fits together to make sense of life.

Robert A. Laidlaw put accepting the existence of God in proper perspective when he wrote, "God exists whether or not men may choose to believe in Him. The reason why many people do not believe in God is not so much that it is intellectually impossible to believe in God, but because belief in God forces that thoughtful person to face the fact that he is accountable to such a God." The problem with the existence of God is not that it is unreasonable. The problem is the fallen nature of man. It is always trying to find a reason not to believe. Why? If God exists, a person cannot be a god unto himself. Blaise Pascal summed it up best, "People almost invariably arrive at their beliefs not on the basis of proof but on the basis of what they find attractive."

That is not true of every person. There are those who truly want to believe God exists. If you ask people the question, "If God exists, would you want to know Him?," you will find most of them will answer **yes**. A recent Harris poll found that 72% of those questioned believe in God. Most people want to believe that there is more to life than what they have experienced up to the present time in their lives.

The good news is there is far more to life than what this world can offer. That **more** can be found through belief in the existence of God. The good news also is that it takes more faith not to believe in God than it takes to believe in God. As I have shown in this book, belief in the existence of God is far more reasonable than to not believe. For one thing, the Bible has stood the test of time as a reliable source in showing the historical and prophetic validity of its content. To not believe in God, a person has to throw logic out the window and believe that something can actually come from nothing. That is absurd. All we have to do is observe the creation all around us using proven scientific methods to come to the conclusion that design, life, intelligence, and predictability all require an absolute Cause.

But more than any of these evidences, our very own being requires a reason that is greater than ourselves that longs for the validation of our existence to have meaning. Just as every child does not feel complete when not having a father, every human being does not feel complete without knowing their Creator in whose image they are made. We come to completeness when we find we truly have a Father who gave us life and loves us. Paul wrote to the Roman Christians, "The Spirit Himself bears witness with our Spirit that we are the children of God." In Jesus Christ we have a Father who is the reason why we exist. We exist because our Father exists.

My Search for Truth

Dark, dark were the recesses of my soul

Emptiness without meaning had taken its toll

Why am I here? What can I know?

No absolutes had left a large hole

Then came the day dispelling the night

When the truth began to rise shining so bright

The darkness fled to escape from the light

And things unseen came into sight

Those things were always there

Though hidden by my despair

Things like proofs of truth to be shared

Things like God exists and that He cares

As my mind was opened to logic and reason

My thoughts entered into a whole new season

From winter to spring I began to ease in

Leaving behind old thoughts that were treason

The old had passed on the new had come

My mind was clear no longer to run

My empty soul had found faith in God's Son

My search for the truth was finally done

By Steve Kern 3/31/2016

SOURCES

∞

Comfort, Ray, 2001, *Scientific Facts in the Bible*, Bridge-Logos, Alachua, Florida.

DeYoung, Don, 2005, *Thousands ... Not Billions*, Master Books, Green Forest, Arkansas.

Geisler, Norman L., Turek, Frank, 2004, *I Don't Have Enough Faith to Be an Atheist*, Crossway Books, Wheaton, Illinois.

Kern, Steve, 2007, *No Other Gods*, Kern Enterprises, Oklahoma City, Oklahoma.

Lyle, Jason, 2009, *The Ultimate Proof of Creation*, Master Books, Green Forest, Arkansas.

McDowell, Josh, 1999, *New Evidence that Demands a Verdict*, Thomas Nelson, Nashville, Tennessee.

McDowell, Josh, Sean, 2009, *More Than a Carpenter*, Tyndale House, Colorado Springs, Colorado.

Morris, Henry M., 1976, *The Genesis Record*, Baker Book House, Grand Rapids, Michigan.

Morris, Henry M., 1989, *The Long War Against God*, Baker Book House, Grand Rapids, Michigan.

Morris, Henry M., 1974, *Scientific Creationism*, Creation-Life Publishers, San Diego, California.

Sproul, R. C., 2003, *Defending Your Faith*, Crossway, Wheaton, Illinois.

Taylor, Ian T., 1996, *In the Minds of Men*, TFE Publishing, Minneapolis, Minnesota.

Turek, Frank, 2014, *Stealing From God*, NavPress/Tyndale House, Colorado Springs, Colorado.

Vine, Unger, White, 1985, *Vine's Expository Dictionary of Biblical Words*, Thomas Nelson, Nashville, Tennessee.

BOOKS BY DR. STEVE KERN

JUDGMENTS GREATEST QUESTION

Dr. Kern's testimony and philosophy of ministry
as the pastor of an inner city church in Oklahoma
City, Oklahoma. **The book is based on the fast
God has chosen in Isaiah.**

EDEN'S VEIL

This is the first book in a three part trilogy.
It is an adventure novel written in the context
of the preflood environment based on a literal
creation interpretation. It is the story of a man
seeking truth in a fallen evil world. He is a
descendent of Cain in search of the Garden of
Eden. The introduction gives an overview of the
preflood. **It gives an alternative to evolution
explanations of how the world began.**

EDEN'S SON

This is the second book in the three part trilogy. After finding the garden through a great deal of hardship the hero of the book fins peace and meaning. **This book tells the story of his adventures and hardships he faces as he returns to tell his people about the one true God.**

EDEN'S TEARS

This is the third book in the three part trilogy. It tells the story of the hero in his later years and what the world was like leading up to the flood. The story transitions to Noah and his family as they build the ark and finally spend the year on the ark. **The book captures many of the possible difficulties Noah and his family faced during that time.**

NO OTHER GODS

Dr. Kern's signature book on creation. It is a verse by verse commentary on the first 11 chapters in Genesis. **It covers the creation, the fall,the flood, and the tower of Babel using science, archeology, and other apologetic arguments to show a literal, historical interpretation is the proper interpretation.**

GOD'S ANSWER TO THE QUESTION OF EVIL

The existence of evil is one of the major
arguments used by anti God detractors to justify
their atheism. **This book gives a clear explanation
why belief in God and the existence of evil
is the more reasonable understanding
of reality.**

THE SIX DAYS OF CREATION

A day to day commentary on the six days
of creation week and the seventh day of rest.
Dr. Kern wrote this in partnership with
Dr. Norbert Smith, an accomplished author
and professor of zoology. **This is an insightful
study given from a theologian and a scientist
who accept the literal creation interpretation
of Genesis.**

GOD'S PURPOSES FOR MARRIAGE

Marriage was established by God
in Genesis 1 and 2. **This book gives an
explanation of God's Purposes for marriage
embedded in the Genesis account.** It covers
expressing the image of God, expressing the
majesty of God, producing the kingdom of God,
and perfecting the worship of God.
The objective is to help married couples bring
glory to God through their marriage.

GENUINE CHRISTIANITY

Many Christians deal with being secure in their faith. They ask themselves "How do I know for sure that I am a Christian?" **This book answers that question and challenges believers to fully embrace the leadership of the Holy Spirit's leadership in their lives.**

www.ingramcontent.com/pod-product-compliance
Lightning Source LLC
Chambersburg PA
CBHW060708030426
42337CB00017B/2796

9780979866777